Reginald Harrison

Lectures at St. Peter's in 1890

Some Urinary Disorders Connected with the Bladder, Prostate, and

Urethra

Reginald Harrison

Lectures at St. Peter's in 1890
Some Urinary Disorders Connected with the Bladder, Prostate, and Urethra

ISBN/EAN: 9783744692090

Printed in Europe, USA, Canada, Australia, Japan

Cover: Foto ©berggeist007 / pixelio.de

More available books at **www.hansebooks.com**

ON SOME URINARY DISORDERS.

LECTURES AT ST. PETER'S (IN 1890).

ON SOME

URINARY DISORDERS

CONNECTED WITH

THE BLADDER, PROSTATE, AND URETHRA.

BY

REGINALD HARRISON, F.R.C.S.,

ONE OF THE SURGEONS TO THE HOSPITAL; HUNTERIAN PROFESSOR OF
PATHOLOGY AND SURGERY AT THE ROYAL COLLEGE OF SURGEONS OF
ENGLAND; VICE-PRESIDENT OF THE MEDICAL SOCIETY OF
LONDON; HONORARY MEMBER OF THE MEDICAL
SOCIETY OF THE STATE OF NEW YORK.

LONDON:

BAILLIÈRE, TINDALL, & COX, 20 AND 21 KING WILLIAM ST.,
STRAND, W.C.

1890.

CONTENTS.

LECTURE I.

The Prevention and Early Treatment of Prostatic Obstruction.

IN the course of lectures that is before us this month I hope to elaborate certain points connected with the diseases of the urinary organs which appear to me to be worthy of your consideration. It will be my endeavour to make these meetings gatherings for clinical objects rather than for the purpose of attempting any systematic description of the subjects to which they relate. With this view I shall be happy as occasion offers, to supplement what I may have to say here by the ordinary work of the hospital as seen in the wards, and on my out-patient days.

Without occupying your time by any further prefatory remarks, I will proceed with the subject selected for to-day's consideration, viz., the Prevention and Early Treatment of Prostatic Obstruction. And I must ask you to notice in the first place that I do not say the prevention and treatment of the enlarged or hypertrophied prostate, as I purposely confine myself to dealing with almost the only symptom, and that the most pressing one, which brings the enlarged prostate under the notice of the medical practitioner. Enlargement of the prostate to some degree is, I might say, the common lot of by far the greater majority of males who have attained sixty years of age or upwards, being, I believe, far more frequent than what at first sight we might perhaps be disposed to admit. Not very long ago for the purpose of ascertaining this point I took the trouble to examine consecutively 100 males who had reached their sixtieth year or upwards, and out of these three-fourths presented undoubted signs by rectal examination of this change. Of the number thus affected not one-fifth of them appeared in any way inconvenienced thereby, some to so slight an extent as not to complain. Nor are there grounds for asserting that because a person has an enlarged prostate a time will come, should he live long enough, when bladder trouble must necessarily supervene. I have watched many men with well marked signs of an enlarged prostate for a number of years, and up to the time of their deaths, either by natural decay or by other diseases, without knowing them to be conscious of any inability so far as their urinary organs were concerned. But though persons may, either by good care, or good fortune, obtain an immunity from those troubles which prostatic enlargement not unfrequently brings with it, there can be no doubt that they carry with them to the end of their lives the liability to suffer in a way which does not apply to those who have not undergone this structural change, a liability which may show itself without any sufficient notice or warning. It not unfrequently happens, as we all know, that a sudden attack of retention of urine, probably traceable to some definite cause which might have been avoided, is the first indication that an elderly man has an enlarged prostate. As a matter of practice it is as an obstacle to micturition that the prostate comes under our notice as surgeons, and this for the most part shows itself, at all events to persons accustomed to observe, by gradual indications rather than by the sudden and unwarned cessation of the physical function of micturition. In far the larger number of cases, even in those where apparently the urinary breakdown is sudden and unexpected, and where the prostate eventually proves to be a serious cause of obstruction the process is a gradual one, and is not unlike what is commonly observed in connection with other obstructive diseases of the urinary apparatus. Hence I consider that with our present knowledge, and in face of the facts to which I have briefly referred, we, as practitioners will do best by approaching the subject in relation to obstruction rather than in reference to a mere increment in bulk of which the individuals may never be conscious.

These cases, however, of much enlarged prostates unattended by symptoms of such a kind as to constitute disease, have been of great interest to me, and the study of them led me some years ago to suggest certain lines of action in practice which seem to me to have yielded some excellent results.

In a number of specimens of this kind it was impossible not to notice that the form assumed by the prostatic growth relative to the outlet from the bladder exercised a considerable influence in determining the presence or absence of symptoms of urinary obstruction. In some forms of enlarged prostate, though the growth was not excessive, the obstacle to micturition was most complete, whilst in others, though the mass attained a very considerable size, it was quite apparent how it happened that the expulsion of urine from the bladder was so little interfered with. By the kindness of my friend, Professor A. Barron, I am able to bring

under your notice to-day an example about which I
attach considerable importance in reference to this
special point I am now addressing you upon, and I
attach this importance to it chiefly on two grounds (1)
because it illustrates how it is that the process of
micturition need not necessarily be interfered with by
the growth of the prostate, and (2) because it shows a
natural adaptation of the parts which I shall endeavour
to demonstrate to you, is capable of being imitated.

The specimen you will see is one of considerable
enlargement of the prostate ; it is taken from a middle-
aged man about whose exact age there seems to be some
doubt, but who probably was rather under than over that
period of life when this change usually commences. He
was under observation for eleven days before he died ; he
had never suffered from any difficulty in urinating,
though the act of micturition was more frequently performed than is usual with persons at his time of life. If
we look at the specimen, we can at once see how it was
that, in spite of the general enlargement, there was no
marked impediment to the escape of urine along the
urethra. You will notice that what is known by the
somewhat incorrect term, as the middle lobe, is deeply
grooved in the centre so as to give a bi-lobed appearance
to the part. Having frequently noticed this process of
channelling or grooving in connection with those cases
of enlarged prostates where micturition was not interfered with, in 1881 I drew attention to this point with
the view of showing that it was capable of being artificially produced with permanent advantage to the patient.

About that time a case came under my notice which
seemed to confirm the views I was then putting into
practice. It was that of an elderly gentleman, well-
known to me, who, in the belief that he had an enlarging

prostate, of his own will and idea took means which seemed to him would be likely to prevent the occurrence of retention of urine. Long before I knew him he had been told that his slight urinary troubles were the early indications of an obstructing prostate, and that he might find himself on some occasion unable to micturate. This so alarmed him that for many years, and almost up to the time of his death at an advanced age, and from other causes, he never allowed a day to go by without passing a full-sized bougie for himself. He remained quite free from any urinary inconvenience, a circumstance which was attributed by him to the means he had adopted. I made a post-mortem examination of him, and found that though the middle lobe of his prostate was considerably enlarged, the level and distensibility of the prostatic urethra were in no way altered, the growth or enlargement being deeply grooved. It appeared, to use the phrase of my patient, that the "maintenance of the water way" was directly traceable to the persistent catheterism that had been most gently and effectually performed. In the further extension of this practice my object has been to mould the growing prostate, and not to endeavour to dilate it, or to procure its absorption as had previously been done by Mercier, Physick, Skey, and other surgeons.

When there is evidence that the prostate, by its increased bulk, is beginning to obstruct micturition, when we endeavour to supplement an imperfect action of the bladder by repeated trials and failures, or when the residual urine which in some degree is nearly always a concomitant with an enlarged prostate is seriously harmful to the individuals either by its quantity or quality, then mechanical treatment in the direction I have indicated may often be resorted to with great and permanent advantage. But there are other grounds except those immediately connected with the function of the bladder, which seem to force upon us the importance of the mechanical treatment of prostatic obstruction at a far earlier date in its history than has hitherto been generally practised. Take the evidence that is afforded by the state of the urinary organs above, namely, the bladder, the ureters, and the kidneys, as showing what the back pressure of the urine in some cases of prostatic obstruction is capable of effecting. In the post-mortem record of the specimen, I have just shown you, where during life there was no reason for suspecting that any appreciable degree of obstruction existed, it is stated "the kidneys were fibroid and atrophied, the two weighing ten ounces, many small subcapsular cysts, early hydronephrosis, no pyelitis. Ureters dilated, bladder dilated and fasciculated." Supposing a surgeon accustomed to see the pathological effects connected with urethral stricture had a specimen of this kind before him in an instance of the latter disorder, I can imagine him saying to his class "here we have an illustration of the enormous and fatal damage that urine pressure is capable of exercising on the important parts above a stricture; these results are in their nature entirely due to mechanical causes, and might have been prevented had means been taken sufficiently early."

Apart, however, from this consideration, it is important to notice what an unfortunate condition of the organs is presented should in the course of the prostatic disorder retention of urine attended with cystitis, as is often the case, supervene. We have the whole of the urinary apparatus from the kidneys downwards laid open as it were to the influence of an inflammatory attack of a most putrid character. Can we wonder that what is commonly called a surgical kidney is one of the most natural results?

But though the more remote consequences of prostatic obstruction are for the most part analogous with those observed in connection with urethral stricture, little seems to have been done to prevent their occurrence. This appears to have been due in a large measure to our failing to recognise the possibility of maintaining in spite of overgrowth a prostatic channel for the escape of urine from the bladder quite as efficient, as I have endeavoured to show, nature herself frequently provides, and which is not liable to that kind of contraction which nearly all strictures constantly undergo. In the next place our forefathers were ill provided with instruments adapted for such purposes, the old metal long curved catheter being almost always associated with any prostatic disorder. I do not think we fully appreciate the great improvements that have taken place within comparatively recent years in the construction of these appliances, and the comfort that has thus been conferred on those requiring such assistance.

Frequency of micturition may oftener be regarded as a sign that the prostate is commencing to obstruct, than any other symptom. As with a urethral stricture a person by the exercise of muscular force will often maintain a very fair stream of urine, though the dimensions of the canal are considerably reduced, so will an individual with a large prostate compensate to a certain point for efficiency by a frequent performance of the act of micturition. Hence it behoves us in all cases where there is an absence of other causes of irritation, such as are presented by stone, tumour, inflammation, or the condition of the urine, to regard frequent micturition in an elderly male, as an indication that he is probably suffering from an early form of prostatic obstruction. The presence of this symptom in an inconvenient degree and in the absence of any other cause such as I have indicated, is sufficient to warrant a physical examination of the suspected part. The discovery of some enlargement of the prostate by exploration of the finger in the rectum is *prima facie* evidence that the frequency in micturition is merely an indication that the growth is commencing to interfere with the proper performance of micturition. Such being the case, it is hardly necessary to point out that the artificial rest to the part which is frequently obtained by the use of opium and belladonna suppositories, except so far as removing spasm is concerned, is really an interference with an adaptation in the manner of conducting the process of micturition which is essentially associated with the altered physical condition of the parts. If in a system of water works you materially alter the shape, and structure of the main valve at the outlet, you can hardly expect that the discharge will be the same as under the previous *regime*.

And now I will proceed to show you by what means we may contrive to improve matters when there is evidence by the symptoms it provokes that the enlarging prostate is assuming a shape where urination is, and will probably continue to be, increasingly obstructed. Having drawn my first illustration from what nature will spontaneously do in reference to this difficulty, I will take a second one from the same source before I proceed to notice the mechanical contrivances which may, under such circumstances, be used with advantage. Persons with large prostates and who are not inconvenienced by them are quite unconscious of the spontaneous process by which the conservation of their urinating powers is preserved. It is not by a rapid destruction or alteration in tissue consequent on some violent inflammation where, as we sometimes see, a permanent advantage follows, but by the induction of a process of which we are absolutely unconscious, and this fact must be uppermost in our minds, so that we may so guide our efforts as nearly as possible to imitate that which nature does not only with impunity, but with so great and permanent advantage.

With the object I have now in view, I would in the first place say that bougies are, as a rule, to be used and not catheters. I am quite sure that unless there is some special reason for it, it is a great mistake to interfere with that degree of residual urine which is met with in most cases of prostatic obstruction, especially in its early form. What we have to do is to bring about such a conformation of the outlet from the bladder as to render residual urine unnecessary. Apart, however, from this point, which is an important one, it must be remembered that a flexible bougie generally enters the bladder much more easily than any catheter, as the stiffening which the eyes of the latter entail render it less capable of adapting itself to the canal. The special instruments I employ for moulding the prostatic urethra are of three kinds—the whip bougie, the ordinary olive-headed instrument, and the bougies *à ventre*, or bellied bougies, which I first introduced some years ago for this purpose. I prefer the last-mentioned, (Fig 1.) in reference to which I can only endorse what I wrote of them nine years ago. (*a*) Whatever instrument is selected care should be taken to ascertain that it can be used easily and without causing pain, as the object in view would thereby be defeated.

(*a*) "The Prevention of Stricture and Prostatic Obstruc on." 1881.

14

The narration of a single case of some years' standing will enable me to complete the particulars which seem to me to be of importance in connection with the modelling treatment of prostatic obstruction. In 1883 a medical man, æt. 62, consulted me for symptoms which he was inclined to attribute to a stone in the bladder. For some months he had been suffering from extreme irritability both by night and day, and particularly in the morning on rising. He was engaged in an active life, and the urgent desire to micturate every two or three hours became extremely irksome and inconvenient. He had a great dislike to any instrumental interference, and up to the time of my seeing him had refrained to use a catheter, although he had been frequently urged to do so. His habits and modes of life had always been extremely temperate and regular. Examination by the finger in the rectum showed that he had a moderately enlarged prostate. As he wished to be searched for stone the introduction of the sound showed by the manner in which the handle of the instrument had to be depressed to enter the bladder that the floor of the prostate was very considerably elevated and enlarged. There was no stone nor evidence of any disease except the enlarged prostate. There were about two ounces or rather more of healthy residual urine. Having explained the nature of the case to him I advised him not to use a catheter, but to accustom himself to the regular in-

FIG. I.

troduction of a suitable bougie. To commence with I gave him one of the whip bougies, which I was then introducing into practice, and asked him to pass it every second or third night on going to bed. In the course of three weeks he got quite accustomed to this very simple form of instrumentation, and I then substituted the smallest sized bougie *à ventre* or prostatic dilator as previously described. He passed this for himself on going to bed about twice a week, and very gradually increased the size of the instrument by three or four numbers. He reported himself at the end of six months, having persistently carried out this treatment in conjunction with the pursuit of his ordinary business. His condition then was as follows:—His propulsive power in urinating had increased, the irritability of the bladder had gradually subsided, the calls to urinate being hardly more frequent than normal, and the amount of residual urine had diminished from rather over two ounces to a little less than four drachms. Nothing else had been done medicinally except the treatment I have mentioned. I saw this gentleman casually two years afterwards, and learnt that he was as well as ever. He told me that he did not think there was the least necessity for using his instrument, but as it was no inconvenience to him he preferred doing so once a week just before going to bed. His urine was acid, and in other respects healthy, and the bladder completely emptied itself.

A useful commentary on this case was suggested to me by the remarks of another medical man to whom I was relating it. "How is it," he said, "that I have not derived a similar benefit? I had to commence the catheter life fifteen years ago, and a day has never gone by since without my passing into my bladder a No. 10 English flexible catheter at least twice in the twenty-four hours." **My reply was this :** " You did not **commence** to draw off your urine until atony of the bladder had proceeded so far that you could not get on without this expedient, and when by the regular and persistent use of your catheter **nature found** out that muscle was not necessary for your act of micturition, either in part or in whole, she ceased to supply it. I **have not** the least doubt that my friend whose **case I have just** related would have been in a similar position to yours had we in the first instance supplied him with a catheter **instead** of a bougie, and told him to go **on** using it."

Before closing this lecture I should like to make a few **remarks in reference** to the larger class of cases, namely, those who **have** hypertrophied prostates but are in no **sense inconvenienced by them. As** I have said before, a person **so** circumstanced must be regarded as more liable to some sudden urinary disturbance than one who is not. Hence I have been in the habit of enumerating **a few** points which in the management **of** elderly males have proved of service.

1. **To avoid being** placed under circumstances when the **bladder** cannot be emptied at will. Nothing is so bad for **a** large prostate, though it may be working satisfactorily as an enforced retention. It is often **the first** cause of a permanent atony.

2. To avoid checking **perspiration** by exposure to cold, **and thus throwing** additional work **on** the kidneys. In **climates like our** own elderly persons should, both in summer **and winter, wear** flannel next the skin.

3. **To be sparing of wines** and of spirits (if used at all), exercising a **marked** diuretic effect either by their **quantity** or quality, select **those which** promote digestion without palpably affecting **the urinary organs.** A glass of hot gin **and** water, or a potent **dose of** sweet spirits of

nitre will not do anything to remove the residual urine behind an enlarged prostate.

4. To be tolerably constant in the quantity of fluids daily consumed. As we grow older our urinary organs become less capable of adapting themselves to extreme variations in excretion. Therefore, it is desirable to keep to that average daily consumption of fluids which experience shows to be sufficient and necessary. How often has some festive occasion where the average quantity of fluid daily consumed has been largely exceeded, led to the over-distension of a bladder long hovering between competency and incompetency. The retention thus occasioned by suspending the power of the bladder has frequently been the first direct step towards establishing a permanent, if not a fatal, condition of atony or paralysis of this organ.

5. It is important that from time to time the reaction of the urine should be noted. When it becomes alkaline or offensive the use of the catheter may be necessary. When a catheter is required it is most important that its selection should not be entirely left to the instrument maker. There are other points to be considered beyond the fact that it is to serve as an artificial outlet for the urine from the bladder. An unsuitable cathether in a prostatic case may do much permanent harm.

6. Some regularity as to the time of performing micturition should be inculcated. We recognise the importance of this in securing a regular and healthy action of the bowels, and though the conditions are not precisely analogous, yet a corresponding advantage will be derived from carrying out the same principle in regard to micturition.

Of the medicines I have found most useful in

conjunction with mechanical means in restoring the tone of the bladder, I would mention here the ergot of rye which I generally give in the form of the fluid extract in cinnamon water. Its value in promoting muscular contractility is generally recognised. In my next lecture I shall consider the treatment of some advanced forms of prostatic obstruction.

LECTURE II.

The Operative Treatment of Advanced Forms of Prostatic Obstruction.

I HAVE selected this subject for our consideration this afternoon, and I would again remind you that I confine myself to dealing with the single symptom of obstruction to micturition, quite independent of any size that the prostate may assume, knowing well that hypertrophy may occur without causing the individual such an amount of inconvenience as to demand surgical interference.

As I have said before, obstruction to micturition may be a process of gradually increasing difficulty, or one suddenly brought about by some casual circumstance which may or may not have been avoided. Whatever may be the mode of its incidence we frequently meet with persons whose function of micturition is entirely and permanently spoilt by the difficulties and complications attendant upon a large prostate. So long as a person so circumstanced can keep himself comfortable by the use of a catheter, so long as the calls for this instrument are not of such frequent occurrence, both by day and night, as to seriously interfere with his health, or to prevent him earning his subsistence, I do not think that

c

further measures are likely to be called for, unless under very exceptional circumstances. Surgery, however, has to provide not only for the ordinary course that a disease is likely to take, but for all the contingencies to which it is liable. Hence I propose to consider in the first place the conditions to which catheterism alone is inapplicable as failing to give sufficient relief, and secondly, the varieties of relief for such states that surgery offers.

I do not think there is any phase connected with advanced forms of prostatic disease which is more intolerable and wearing to the individual, than that ceaseless desire to evacuate the bladder which is occasionally met with. Yet in practice, instances will come under your notice where the whole life of the individual both by day and night, is as it were given up to the single act of introducing the catheter. Here the instrument, as a patient once said to me, becomes a Tantalus of the worst form, as not the slightest relief follows its application. Then there are cases where the mechanical act of introducing the catheter is an increasingly difficult one by reason of the obstacle that the growing prostate presents. A patient who has been using a catheter for some time with great advantage finds that he has to select a smaller size, then he tries a different shape, and so he goes on until at last both he and his medical attendant discover that there is always difficulty, whatever kind or size of catheter is used. Then hæmorrhage attends these trials, at first occasionally and afterwards constantly, and this is frequently followed by cystitis and other complications, amongst which orchitis may be mentioned. This condition of affairs does not often last long unless some easier mode of evacuating the urine is provided.

In a third class of cases you will find that in the course of its growth a portion of the prostate in the form of one or more pendulous growths, encroaches upon the interior of the bladder and keeps up a constant state of cystitis, often far in excess of what is seen associated with the most irregular stones. Here the urine is generally loaded with a muco-purulent deposit, in which particles of blood will frequently be noticed. Further if you look at specimens of this kind which will be found in any pathological museum, the ureters will be seen to be patent and largely dilated, evidently showing not only how great the obstruction has been, but how easily the damage may enter the interior of the kidney. What can catheterism and washings-out do for cases coming under this category ? Then there are instances of a very exceptional nature where the ordinary modes of treatment are inapplicable. In one case that occurs to me the prostatic urethra became so inordinately sensitive, and the spasm that catheterism excited in the prostatic urethra, so great that the patient actually wore an instrument in his bladder, more or less constantly for over two years. I believe this case was in the first instance one of those rare ones of prostatic fissure, analogous in many respects with what is seen within the anus and the female urethra. I operated upon him three years ago, dividing his prostate, and at the same time removing some secondary phosphatic calculi. He is now alive and well and though he passes his urine more frequently than natural he has never since had occasion to use a catheter. Though in the majority of instances the advanced forms of prostatic obstruction are arrived at by a process of increasing difficulty in urination, the disease sometimes assumes this position without giving anything which could be spoken of as an adequate warning. Retention of urine, following apparently some trivial cause, suddenly occurs and the practitioner finds himself face to

face with one or more of the complications to which I
have just referred. There is great difficulty in introducing or re-introducing the catheter by reason of the form
the prostate assumes ; the retention of an instrument in
the bladder is intolerable to the patient, if it is tied in
it is withdrawn and all the trouble and pain connected
with its re-introduction have to be repeated; or the
engorged condition of the parts is attended with hæmorrhage which recurs every time instrumentation has to be
practised. These are phases of the acute form of obstruction which most practitioners are not only familiar with,
but who know from experience how futile catheterism
alone has proved under such circumstances.

For the relief of these difficulties, whether occurring
gradually or suddenly, various methods of treatment
have been practised, to which I shall now refer. In
dealing with this part of my subject to-day, I hope I
may be pardoned if I do so from a somewhat personal
standpoint, as I think it will be seen that I have taken
a considerable interest, not entirely in a passive form, in
this aspect of the question.

In 1881 I published the particulars of a case where to
meet the pressing difficulties connected with an obstructing prostate such as I have just referred to, I adopted an
expedient which I believe was entirely novel at the time.
I had long felt that the operation of puncturing
the bladder from the rectum for retention of urine in
cases of an enlarged prostate was one that could only be
recommended for a very temporary purpose, inasmuch as
it was impossible to retain the tube sufficiently long in such
a position as to permit of an adequate degree of bladder
drainage being practised with any regard to the comfort
of the patient. Nor did the supra-pubic method of
puncturing the bladder, and retaining a tube in this

position recommend itself to me, chiefly on the grounds that it seemed to me to be neither a convenient spot for the patient, nor was it sufficiently dependent to secure perfect drainage. With these views in my mind, I determined to puncture the bladder, when the necessity arose, from the perinæum through the enlarged prostate, and to retain a drainage tube in this position as long as might be necessary. In the instance I am now referring to, the patient was a man of 84 years of age, who was much reduced by an advanced form of prostatic obstruction. Though a catheter could be introduced, I clearly saw that unless some more efficient and direct method of urine drainage was provided, he would speedily sink. Under these circumstances, I tapped the bladder from the perinæum through the large prostate with the instrument I had prepared for this special purpose, and which I will now show you. The operation of tapping the bladder when distended from this position is one that is not attended with any difficulty. The point of puncture is one inch in front of the anus, and the direction in a line towards the umbilicus. The trocar is so devised that urine escapes from the handle of the instrument immediately the interior of the bladder is reached. In the instance now under notice, the patient was at once relieved, and in twenty-four hours he was up, sitting in an easy chair—an important matter with old people. I will not trouble you with the details of this extremely interesting case as they have already been published in full (a). For about six weeks after the bladder was tapped, the patient passed urine entirely through the prostatic cannula, and he was able to go about as if nothing ailed him, by means of a tube

(a) "Surgical Disorders of the Urinary Organs." Third edition. By Reginald Harrison.

apparatus specially adapted for this purpose. Then it was noted that urine, in gradually increasing quantities began to flow through the natural passage, indicating that the prostate was ceasing to obstruct. The cannula was removed six weeks after the operation—the wound completely closed in the course of a few days, and the patient entirely recovered. He died at the age of 90 without any recurrence of his prostatic troubles. What was most interesting to observe in this case was the prostatic atrophy that seemed to follow the operation. I examined this patient two years after the operation, and could find no evidence that the prostate was enlarged. I have had on two occasions to repeat the operation, with results entirely corresponding with the case I have related more at length. I have not yet been able to collect the evidence in reference to the shrinking of the prostate which has followed this procedure, but my own observations in reference to this particular point have been corroborated by others. Mr. Weston, of Stafford, has kindly given me the particulars of a case of a very similar nature, where, after the operation, a very considerable diminution in the size of the prostate was observed.

But though this method of proceeding was especially applicable to those extremely rare cases of retention of urine due to a large prostate, where it was thought that no catheter could be introduced, it seemed to me that there were others which would be more advantageously dealt with by an opening from the perinæum, which would allow of digital exploration of the enlargement followed by drainage. This procedure formed the subject of a communication I made before the International Medical Congress at Copenhagen in 1884. It consists in puncturing the membranous urethra on a grooved staff in the median line of the perinæum sufficient to permit the introduction of the index finger within the prostatic urethra, to this extent resembling Cock's operation. Having thus obtained access to the interior of the bladder by a very limited incision, I have dealt with the prostate in three different ways. (1) By a division of the floor of the prostate as a preliminary to the insertion of a drainage tube. (2) By the enucleation or removal of some portion of the prostatic mass before putting in the tube, and (3) by the insertion of the drainage tube alone. I have now operated in these ways on something like thirty-five cases of advanced prostatic obstruction, excluding malignant growths in males over 60 years of age, for reasons such as I have stated, where catheterism proved insufficient to give adequate relief. In eighteen of these cases, either the floor of the prostate was divided, or a portion of the enlarged prostate removed before the drainage tube was introduced, whilst in the remainder the perinæum was opened merely for the introduction of the tube. I should like to say a few words about these operations, which have now ranged over somewhere about seven years, and the points that I would speak to on this occasion are these: (1) The risk of operations of this kind. (2) The present or subsequent condition of the persons, as far as I know, who have been thus treated, and (3) the effects that may have been observed on the enlarged prostate. Of the deaths, I only know of three, one, a week after the operation, and another four weeks, and the third two months afterwards. In these instances the patients, judging from the condition of the kidneys and ureters from previous back-pressure, lived longer and more comfortable than they would otherwise have done. One man, aged 68, died over six months afterwards from a paralytic seizure. During this illness it was not found necessary to use a catheter for him. Two persons died over twelve months after the operation, from causes that I have not been able to ascertain, but as in this interval they had been well and active, they can hardly be included in the list of mortality. What, I would ask, would the mortality have been, had all these cases been left to endure the troubles that attend constant catheterism under difficulties? Of the remaining twenty-nine persons, something is or has been known of nineteen of them, at intervals of time over twelve months after the operation. Three of these have a permanent opening in the position of the wound. Two of these catheterise themselves in this way, and report themselves as perfectly well. The third had, in addition to an obstructing prostate, a tight urethral stricture, and now micturates at natural intervals through the perinæum without the aid of the catheter. Of the sixteen whose artificial openings have soundly closed, ten are quite independent of the use of the catheter, three of them being extremely feeble by reason of advanced years. Of the remaining six, four have to use the catheter as before the operation, but do so without difficulty, or without causing hæmorrhage. The two remaining cases have in addition had temporary difficulty in catheterism. In one of these the obstacle seems to have been due to the impaction of a calculus from which he has occasionally suffered for some years. Of the remaining ten, of whose condition nothing is known or can be ascertained, four are believed to have left England altogether.

And now I come to the third question which a retrospect of these cases suggest, namely, Have any effects on the large prostate been observed? I have already stated that shrinking of the enlarged prostate

has been known to follow when the bladder has been
tapped through it, and the cannula retained for some
time. I am equally certain, and have pointed it out to
others, that I have seen the same shrinkage occur where
the urethra has been opened and a tube introduced for
the purpose of drainage without any damage being
inflicted by section, puncture or otherwise on the prostate. This has been more or less evident in the majority
of the 19 cases previously referred to, in some to a very
marked degree. Hence one is less surprised at finding the
marked and permanent benefit in micturition which some
of the cases referred to undoubtedly presented. This observation has led me to inquire as to the precise circumstance
connected with this class of operations on the enlarged
prostate which seemed to have been coincident with this
extremely important change. The facts appear to point
to the inference that where the bladder is drained incon-
tinently for some weeks, I speak of from three to six
weeks, an enlarged prostate will undergo a palpable
degree of shrinkage, and I make this statement on the
ground that where the drainage has been thus prolonged
the urine will be found to force its way in spite of the
tube along the natural passage. I have noticed this on
so many occasions that when it happens I take it as a
sign that the tube may be removed. That it has proved
a correct signal I have not the least doubt, as healing of
the wound has rapidly followed. Further than this evidence was also afforded me on several instances that the
diminution in size was apparent to the finger from the
rectum. If, as I have ventured to suggest, hypertrophy
of the prostate is directly associated with the large
amount of muscular element that enters into its com-
position, it is not surprising that the induction of a
state of complete physiological rest should be followed

by the result I have noted. However this may be, the fact remains unaltered.

Before referring to the details connected with my method of operating and illustrating it, I should like to briefly refer to the practice of suprapubic prostatectomy which has been recently introduced by Mr. McGill, of Leeds. Taking advantage of the improvements which have been so warmly advocated by Sir Henry Thompson, and others in regard to this way of opening the bladder for stone and other tumours, Mr. McGill has utilised it for the purpose of removing more or less of the hypertrophied prostate. At the Leeds meeting of the British Medical Association last year several patients were exhibited, who certainly derived great and I believe permanent benefit from the operation. When, however, the subjects and the circumstances of suprapubic prostatectomy are taken into consideration, I do not think that it is an operation which is likely to be very extensively employed. In hale elderly persons where the evidence is very conclusive that a prostatic mass is projecting into the bladder and irritating it, we may occasionally avail ourselves of it with advantage. With the old, feeble, and broken-down, who oftener than otherwise furnish the victims of advanced forms of prostatic obstruction, we must be content with measures which, whilst affording immediate and complete relief, entail far less strain on the reduced powers of repair of the individual than is implied by a suprapubic removal of more or less of the prostate. For this reason, especially, in addition to others, I am inclined to limit the operative procedure to the opening and draining of the bladder from the perinæum as affording a mode of relief which, so far as I have observed it, is unattended with risk and is capable of giving not only immediate but permanent relief.

And now I will refer to the details connected with my operation. The patient being placed in the lithotomy position, a staff with a median groove is passed into the bladder. There may be extreme difficulty, from spasm chiefly, in introducing this or any other instrument up to the time the anæsthetic is administered, but when this is done I have always found the obstacle at once disappear. The staff being held in position the perinæum is punctured with a knife one inch in front of the anus, sufficient to enable the index finger of the left hand to be placed on the groove in the staff. No dissection is at all necessary or desirable. If the perinæum is very deep I sometimes clear the staff at the bottom of the puncture with a blunt knife I had made for this purpose. The groove having been bared the small gorget is made to slide along it into the bladder, the handle of the staff being drawn towards the operator to facilitate this movement. Then the finger can be passed within the prostatic urethra. If the latter is felt to be tolerably clear the tube may be at once introduced into the bladder. If the condition of the prostate is such as to offer any obstacle to the passage of the finger or the tube I then introduce a curved probe-pointed knife fairly within the bladder and divide the floor of the prostate directly downwards. This is best done from within outwards. In three instances I have also removed more or less pendulous portions of the growth. The stiff tube is then introduced and retained usually from seven to ten days, then a soft tube is substituted. For the latter purpose I have employed a contrivance suggested by Mr. Annandale, which enables the patient to get up and go about, as shown in the figure (Fig. 2.) The soft tube which is tucked into the waist belt is connected with the perineal tube, so that when the patient de-

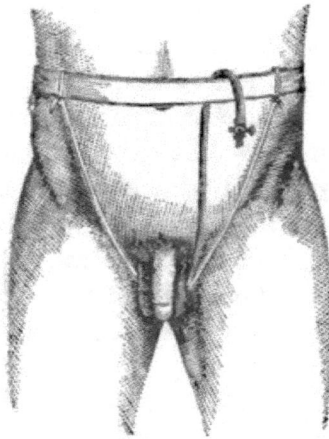

FIG. 2.

sires to pass water all he has to do is to let down the former, and turn a tap which is attached to it. Drainage has thus been carried out in the instances I have already referred to from ten days to six weeks. When it is found that in spite of the tube the urine forces itself, as it will do, along the whole length of the urethra, this is a signal that the drainage may be discontinued and the opening allowed to close. The time the tube should otherwise be retained depends of course on the conditions that each case presents. So long as the urine remains alkaline and offensive so long must it be utilised, not only for drainage but as the most convenient way for washing out the bladder. For the latter purpose I find in the first instance a solution of boracic acid, a drachm to a pint of tepid water, most serviceable. If under the use of this the urine does not become normal I usually substitute the neutral sulphate of quinine in the proportion of two grains to the ounce of water. Creolin has recently been used for a similar purpose, but I have not yet tried it sufficiently often as to be able to speak to its virtues. I will conclude this lecture with the narration of one of my earliest and also one of my most recent cases.

The first was a man, æt. 73, in February, 1884, when he first came under my notice. For some months he had had to practice catheterism for an enlarged prostate with a considerable amount of residual urine. After a time there was increased difficulty connected with the introduction of the catheter, and serious hæmorrhage frequently took place. His desire to urinate and his strainings were incessant, and nothing, after three weeks trial with various kinds of instruments, appeared to give him any relief. In February, 1884, he underwent perineal prostatotomy, and wore one of my tubes continuously for seven weeks. The wound closed within a few days after drainage was discontinued. I saw him at intervals for two years after the operation. He remained quite well, and never had occasion to resort to the catheter again. On several occasions I passed an instrument for him, and could find no obstruction in the prostatic urethra, nor did his residual urine exceed a few drachms. This man was rapidly sinking under the irritation, caused by the constant use of the catheter, and there was no doubt in my mind that his improvement was due to the drainage that was practised.

The most recent case was that of a man, æt. 75, whom I saw in consultation with Mr. Woodforde, and who subsequently was admitted into St. Peter's Hospital for treatment. The history and particulars of the case were as follows :—He had had retention of urine and frequency in micturition for eighteen months. For the past six months the desire to urinate had been constant, and was practically unrelievable by the catheter. When I first saw him he was using the instrument every hour both by day and night, the urine being most offensive, and loaded with pus and mucus. As various means for washing out the bladder had been tried I suggested his being admitted into hospital. Shortly after his admission into St. Peter's I introduced a grooved staff and opened the perinæum for the purpose of putting a drainage tube into the bladder, as I have already described. The tube was worn for a week in spite of a good deal of prostatic spasm, when he was able to return to his home. The stiff tube was then replaced by the softer apparatus. Unusual difficulty was experienced in wearing the latter in consequence of the severe spasms that occurred. In time, however, this subsided, the condition of the urine improved and the wound speedily healed on the removal of the tube. In spite of the difficulty connected with the retention of the drainage tube, the only instance where in my practice this has occurred, the man's condition is now (six months after the operation) as follows : —In the daytime he can get about and follow his employment without having to use a catheter. During the night he passes his catheter twice, and sometimes thrice, as he considers that it is an advantage to him, though I have no doubt he could dispense with it. On each occasion I have examined it his urine has been normal. There can be no doubt that though the circumstances connected with the operation were not as favourable for success as I could have wished, this patient's condition since the operation is a greatly, and, I believe, a permanently improved one.

LECTURE III.

Points in the Therapeutics and Hygiene of the Bladder.

As in other directions, I shall hope to show that progress has been made, not only in reference to the more strictly operative measures connected with the urinary organs, but also in all that relates to the therapeutics and hygiene of this class of affections.

In comparing the various systems of which the human body is built up, such for instance as the nervous, respiratory, circulatory, digestive, and urinary, in addition to others that might be mentioned, it appears to me that the urinary more especially enjoys a condition for the action of drugs upon it which is not equally shared by the other contributing systems, except perhaps by the digestive. For, as I shall presently have to demonstrate to you, not perhaps chemically, but at all events clinically, we apparently derive much of our force relative to drug action on the urinary apparatus by the direct control of certain agencies which it is the special province of the kidneys to eliminate, in addition to those general effects upon the body as a whole which the particular drug in question has the power of exercising.

The physiological action of drugs on the urinary apparatus in its general and special application is far too large a subject for me to approach on an occasion such as this, even if my limited knowledge of it as a practical surgeon permitted me under any circumstances to undertake a discussion of this kind. I must therefore limit myself to one aspect of the question with the view of pointing out certain properties of normal urine relative to the tissues with which it comes in contact, the effects which are thus produced, and how the latter may be modified for surgical purposes.

That normal urine possesses by the decomposition of its urea the power of acute tissue destruction is evidenced by those cases of extravasation of urine which most frequently occur in connection with injuries to the urinary organs, and the advanced forms of organic stricture. Here may be seen the rapid conversion of the whole area, which comes in contact with urine so effused into a mortifying mass. This was so well illustrated in a case of an extremely rare nature which recently came under my notice that I will give you the particulars of it.

It was that of a boy, æt. 14, who I saw in consultation a few weeks ago. His history was as follows : he had suffered for years from an irritable bladder, and pain following urination, but nothing appears to have been done for him, his parents not wishing an examination to be made. Four days before I saw him, when playing with some friends in the garden, he fell heavily on his perinæum astride over some palings, and then on to his side. This was followed by swelling of the supra-pubic region, extending to the penis and scrotum. There was no discharge of blood at the time of the accident from the urethra, nor was retention of urine complete. When I saw him four days afterwards the supra-pubic region and the skin of the penis were gangrenous, and the scrotum distended. The perinæum was flat and not œdematous as is usually seen when rupture of the deep urethra from injury occurs. I could not pass a catheter simply for the reason that the glans penis was entirely concealed by a large œdematous and sloughing prepuce. As it was quite clear that extravasation of urine had occurred under somewhat unusual circumstances the patient was put under chloroform. I then proceeded to deal with the different points where relief was obviously required to let urine and disorganised tissue escape, this consisted in making the incision as for supra-pubic cystotomy, incising on either side the raphè the distended scrotum, and slitting up the sloughing prepuce along the dorsum of the penis where the tension was greatest. This enabled me to discover the glans penis, and to introduce a silver catheter into the bladder. This at once led to the detection of a stone which seemed to be held within the prostatic urethra. Though not provided with instruments for lithotomy I thought the stone should be removed at once. I therefore cut down in the median line on the silver catheter, and opened the urethra in the membranous portion. This enabled me to feel the stone with my index finger, though as soon as I touched it, it fell back into the bladder, however, with an ordinary pair of dressing forceps, and my finger in the rectum I succeeded in extracting a pointed stone of a torpedo shape, weighing nearly three drachms. It was composed of light coloured uric acid with a coating of phosphates, and is represented in the figure (Fig. 3).

From the examination I made of the parts I have not the slightest doubt the stone penetrated the bladder immediately behind and below the pubes at a point corresponding with what is commonly called the neck of the

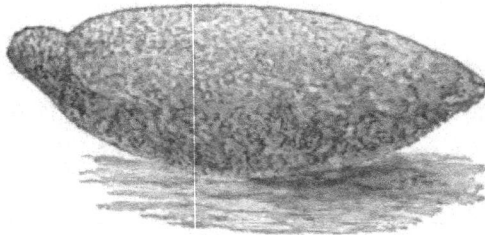

Fig. 3.

bladder. This permitted urine to be effused into the space known as the porta vesica of Retzius, and occasioned all the disastrous effects I have mentioned. The free drainage which the lithotomy wound provided was an important factor in the progress of the case. Though there was considerable sloughing of the prepuce and skin of the penis, scrotum, and supra-pubic region the patient made a rapid recovery. The case is unique, so far as my experience goes, for though instances have come under my notice where calculi have made their way through the perinæum by a slow process of inflammation and suppuration, I am not aware of an instance where the bladder has been suddenly ruptured by a stone it contained in the way I have described.

. Cases, such as these, however, may not only be used for illustrating the acutely destructive properties of urine, but as indicating how completely Nature provides for all the contingencies of life. What could only happen to a man who was unfortunate enough to rupture his urethra when removed from surgical aid unless the urine contained within itself

a property for effecting its own escape? I am now seeing
a patient for a perineal fistula whose life was undoubtedly
saved in this way. He was travelling in the bush last
year in South Africa, and had the misfortune to rupture
his urethra by his horse falling when he was hundreds of
miles away from proper medical assistance. Acute suppuration in the perinæum and sloughing of the scrotum
followed, but in spite of this he recovered. He is now
under treatment for three fistulæ through which the
whole of the urine was discharged for many months
during a process of convalescence under most unfavour-
able conditions. Still, the acute sloughing which followed
the accident, and for which the urine was responsible,
undoubtedly saved the life of the patient.

But though we have evidence in cases such as these
of the destruction certain elements of the urine are
capable of bringing about, we have further proof, if this
were required in instances where the urine by reason of its
altered character is incapable of undergoing that chemical
change which is necessary for acute tissue destruction.

I recorded a case (a) which came under my notice
some years ago, where a middle-aged man who was being
attended for Bright's disease suddenly developed what
was supposed to be an œdema of the scrotum. The
case proved to be one of urethral stricture, complicat-
ing Bright's disease, and causing acute scrotal extravasation of urine. But though the swelling had been in
existence for twenty-four hours before I saw him it seemed
curious that no signs of inflammation and gangrene appeared imminent. However, as the tension was considerable I incised the parts involved in the extravasation.
As the fluid escaped from the incisions I noticed that it
had not the ammoniacal odour which is always so per-

(a) "Lectures on Urinary Disorders." Third Edition. P. 46.

ceptible in such instances. I was somewhat puzzled for an explanation how it was that extravasated urine failed to cause signs of gangrene. I subsequently collected some of the urine as it trickled through the wounds, and compared it with some drawn off from the bladder by the catheter. I found them identical, and chemical analysis proved that in both there was an almost complete absence of urea, the specific gravity of the urine being as low as 1004. This, then to my mind solved the mystery, and explained that as there was no urea to decompose there was no source for the production of the ammonia by which the destruction of tissues in connection with extravasated normal urine is effected. By the absence of urea the urine was rendered chemically harmless to the tissues with which it came in contact.

Passing, however, from instances of acute tissue poisoning I shall have no difficulty in showing you that the same fluid is capable of being placed under circumstances where certain general septic effects may in various degrees be developed. If any one who is interested in this subject will take the trouble to turn to the records of internal urethrotomy as practised only a few years ago they will find in the rigors, fever and occasional deaths that followed this operation, the most unmistakable evidence that experiment is capable of affording, of some forms of urine poisoning which seemed to be associated with certain conditions under which the urine was placed by the nature of the operation that was practised. I was so much impressed with this at the time that it led me to alter my practice materially in reference to an operation which, on other grounds, had much to commend it.

In 1885 I published (a) a series of cases, where in some extremely bad cases of stricture unfitted for any

(a) *Brit. Med. Journ.*, July 18, 1885.

other kind of treatment, I performed internal urethrotomy, and at the same time provided for thorough urine
drainage by a puncture in the perinæum, through which
a tube was passed. Time will not permit me to enter
into this subject as fully as I could wish, but putting
aside collateral evidence and advantages connected with
the proceeding advanced, I established the facts (1) that
the rigors and fever following operations on the urethra
were entirely septic, and (2) that they could be avoided
by providing for more perfect urine drainage. But
though it may not be possible or expedient to construct
all wounds on such principles as to secure healing without the risk of decomposition and absorption of some of
the element of the process, which is the essence of antiseptic surgery, the practice of Lister has demonstrated
how much may be artificially done by the employment of
antiseptics to secure this end. If the urine, as I have
demonstrated, can be placed relatively to a wound under
conditions such as to make its absorption poisonous to
the system at large, it has, on the other hand been
proved possible to so influence it artificially as to render
such a contingency extremely improbable.

And this stage of the argument leads me to lay stress on
what has been spoken of as the sterilisation of the urine by
artificial expedients. Modern therapeutics have undoubtedly shown that some drugs can so influence normal urine
as to cause it to behave very differently from what it
would otherwise do when brought into contact with certain kinds of wounds. The power of quinine in connection
with operations on the urinary organs has long been recognised, and there can be but little doubt that this is
directly associated with the fact that it is so largely eliminated by the urine. Still more striking is the testi-
mony of Dr. Palmer, of Louisville, U.S.A., who found

that he could so sterilise the urine by the administration
of boracic acid in ten grain doses as to prevent the occurrence of urethral fever after such operations on the
urethra as internal urethrotomy. My own experience
in connection with the latter operation corroborates
this, for after the use of antiseptics, principally through
the medium of the urine, I have enjoyed an immunity
from septic fever attacks after operations on the urinary
passages which compares most favourably with what was
previously observed. Nor am I disposed to think that
the power of sterilising the urine so as to render it in-
nocuous when placed under conditions where otherwise
it would be liable to generate septic influences is limited
to quinine and boracic acid. Further investigations will
no doubt show that many other principles are capable of
exercising somewhat similar powers. For instance, I have
frequently been struck with the action of hypophosphite
of soda in half drachm doses in some purulent affections
of the urinary organs, and I have but little doubt that
this is due to the direct influence of the drug as a bac-
tericide in the way that I have endeavoured to lay stress
upon.

Passing, however, from the acuter forms of local and
general urine poisoning, I will proceed to notice others
belonging to the same category. It is now some years
ago since Sir Andrew Clark drew attention to certain
febrile symptoms occasionally following catheterism,
to which the term catheter fever has been applied.
In the general discussion which followed the publica-
tion of these observations, it appeared to me that insufficient prominence was given to the fact that residual
urine is an absolute necessity to some bladders, and that
if an attempt is made to do without it something worse
is provided—that is to say, septic urine is substituted for

aseptic. In a number of observations I made in reference to this point, I found that in certain cases of enlarged prostate, where the bladder was not generally
atonic, the removal of acid residual urine was followed
by the rapid conversion of subsequent excretions into a
more or less ammoniacal compound loaded with bacteria.
It was under such circumstances that catheter fever
was seen, and I cannot find any evidence that it has
ever been met with where the urine was proved to be
normal.

In the course of the inquiries to which I have just
referred, I found from repeated examinations that it was
not difficult to explain how urine thus became acutely
septic. Where a bladder has been accustomed, we will
say, to contain three or four ounces of urine, the sudden
withdrawal of the latter establishes a condition of
flaccidity of the vesical walls which is favourable to the
admission of air either by the instrument or the urethra ;
in the next place, there is an excessive exudation of
mucus from the prostate and adjacent parts ; and lastly,
sanguineous pouring out takes place from the vesical
vessels which are thus suddenly deprived of their support. In this way material for decomposition is amply
provided. So satisfied am I that the physical condition
of the suddenly and completely emptied bladder has
everything to do with the setting up of so-called
catheter fever—a term which must be regarded as
merely expressive of what other cavities and spaces in
the body are capable of producing whenever their con-
tents are in any way analogous—that I consider very
special precautions should be taken in the first instance,
whenever residual urine has to be artificially and syste-
matically removed, with the view of preventing the
aseptic conditions to which I have referred being pro-

vided. What surgeon does not recognise the necessity of doing so where a chest or a psoas abscess has to be tapped with the view of averting what might, with equal force, be called a trocar fever? Where the urine is acid to commence with, it should be removed by instalments; and where alkaline or unhealthy, it should be replaced by some antiseptic fluid, the quantity of which may from time to time be reduced. There is no better antiseptic for the bladder than healthy urine; there is no more dangerous one than when the latter is made to undergo active decomposition.

In failing to recognise a septic cause for the fever which occasionally accompanies catheterism, and referring it to some occult morbid condition of the nervous system, I cannot help observing that the same kind of explanation was advanced up to a few years ago as explaining the rigors and fever following the use of instruments for urethral stricture, until I demonstrated by a series of operations (a) that these phenomena were due to the absorption of some products of the urine, probably of the nature of alkaloids. On this being recognised and provided against, as it now is in most operations on the urethra, the occurrence of such complications, often attended with considerable risk, has been largely diminished, if not entirely prevented, in the practice of most surgeons.

Having briefly reminded you of certain local and general poisonous effects that normal urine is capable of producing, and how these effects may to a large extent be prevented or provided against, I will proceed to notice certain drugs which seem to me to exercise, probably in the mode in which they are eliminated, a beneficial effect on some of the disorders connected with the urinary

(a) Lettsomian Lectures, 1888.

apparatus. In the present day, when we are overwhelmed with the number of new drugs coming under notice, one is rather disposed to speak with reservation on such a subject. Still, on the other hand, there can be no doubt that the field of therapeutics is steadily extending, and that valuable additions are from' time to' time being made. Early in 1886 my attention was first called to the use of Pichi (*fabiana imbricata*) in connection with some diseases of the urinary tract. The case in point was that of a gentleman who had long suffered from attacks of renal colic and hæmaturia. On several occasions he had passed uric acid stones of considerable size after much pain, and once I had to crush a calculus for him which was too large to escape along the urethra. Whilst travelling in Bolivia that year he was seized with a similar attack, and was treated entirely by means of an infusion of Pichi, which he seems to have drank *ad libitum*. He got through this attack very much better than previous ones, a fact which he attributed to the new medicine. He brought me home some specimens of the raw material such as you see on the table.

For the last four years I have been using this drug in the form of an infusion and fluid extract, in drachm doses of the latter, pretty extensively both in private and hospital practice, with, I believe, considerable benefit. I have found it useful in many instances of the following:—1. In renal colic and the passing of calculi through the kidneys and along the ureters, attended with hæmaturia: though not exercising any solvent power it seems by its action on the tissues in some way to favour the escape of the stone, and thus to suppress the bleeding. 2. In the hæmorrhage which frequently accompanies cancer of the bladder. In the case of a medical man who was recently under my observation it

E

certainly gave more relief than anything else that was
tried, as I have also noted in other instances. Lastly,
the sedative action of the drug on the mucous membrane of the bladder has proved beneficial in many
instances of irritability connected with a large prostate.
After the bladder has been properly cleansed by irrigation and disinfected, it has been frequently found that
the calls to urinate were far less urgent when the Pichi
was being used.

Acting somewhat similarly though less astringent in
its properties, and therefore of less value when there is
hæmorrhage, I have found an extract prepared from the
berries or fruit of the Saw Palmetto (*Serenoa serrulata*)
very serviceable. Writing in reference to his own case, a
surgeon in America says:—"I do not know whether you
have used the saw palmetto in cases of prostate trouble.
I have used it several times with apparent advantage."
It is stated to be a plant of the palm tribe, which is indi-
genous to the coast of Florida. It appears in the first
instance to have been used medicinally as a demulcent
in connection with irritative disorders of the respiratory
mucous membrane. It is also stated to possess fat producing properties in the case of animals that feed upon
its fruit. Messrs. Bell and Co., of Oxford Street, have
prepared me a fluid extract from the berries, which, in
doses of half a drachm to a drachm in water has proved
of service in cases of irritable bladder. It seems to act
something like pareira, and is a good substitute for it.

Of the chemical products which I have tried in this
class of disorders, I would mention saccharin in half
grain doses. Attention was first drawn to it by Dr.
Little (*a*) as being useful in preventing ammoniacal
change in urine in cases of cystitis. Where the mucous

(*a*) *Dublin Journal*, June, 1888.

membrane of the bladder throws off large quantities of mucus and the urine undergoes rapid ammoniacal decomposition, I have on several occasions noted that the latter becomes healthy and acid under the use of saccharin. On discontinuing the drug I have also observed that the urine will speedily return to its original condition. Hence it may be found useful in readily providing against conditions which cannot be radically altered. Dr. Thomas Stephenson and Dr. Woolridge (a) have shown that saccharin may be taken for a considerable period without interfering with the digestive or other functions of the body.

Before closing the lecture I will show you a preparation which was first made for me some years ago by Clay and Abraham, of Liverpool, which I have now used rather extensively. It is described as the borocitrate of magnesia. My attention was first called to it by a paper by Dr. Kochler, of Kosten, (b) who advocated its employment in cases of uric acid calculi and gravel. It is prepared by dissolving a natural borate of magnesia which is found at Strassfurt in citric acid. It forms a white powder with a sourish taste, and is given in teaspoonful doses in a tumbler of warm water two or three times a day. I have tried it in several cases of impacted renal calculi, whfch have come to this hospital with the view of having the stone removed by operation. Here are two specimens of stone which have been passed by two patients within the last few days who had been taking the boracite for some weeks previously for attacks of renal colic and hæmorrhage. One of the stones you will see, presents a slightly worm-eaten appearance as if it had been exposed to some solvent action by which its

(a) *Pharmaceutical Journal*, Dec. 1st, 1883.
(b) *Berlin Klin. Wochen.*, Nov. 3rd, 1879.

loosening and ultimate escape had been facilitated. I do not pretend to offer any reasons based on the chemistry of the subject for what I am showing you, all I can say is that I have frequently known the discharge of these bodies, whose presence had previously been suspected, to take place after the use of the salt. It may be all it does is to secure that the individual shall take at stated times more fluid than perhaps he would otherwise do; an important point upon which Sir William Roberts has laid stress. I am, however, disposed to think, from what I have seen, that it does more than thus induce a person to flush his kidneys with a bland fluid by no means disagreeable to take, but that it is capable of modifying or altering the crystalline form in which uric acid is discharged and of exercising a solvent power on some kinds of urate stones.

LECTURE IV.

HÆMATURIA: ITS SIGNIFICANCE AND SURGICAL TREATMENT.

I HAD intended to bring under your notice to-day some points connected with the selection of catheters and other instruments used in the treatment of urinary diseases ; I find, however, I have already said a good deal upon the subject, and that what remains to be added may be incidentally introduced with other material. I, therefore, propose speaking to you about hæmaturia in more special reference to its surgical significance. In order that I might provide myself with a text, and so approach a very wide and important subject with some degree of system and concentration, and at the same time with a direct bearing upon practice, I have carefully gone over my note books containing the records of my private practice, and selected the last hundred cases of true hæmaturia that have come under my observation. Apart from my notes most of these cases, being comparatively recent, are within my recollection, and in regard to a considerable proportion I can speak from a more continuous knowledge of them than I could do of a similar number of hospital or public patients.

For it must be noted that cases of hæmaturia often run over very considerable periods of time without seriously injuring or inconveniencing the individuals who suffer in this way. I have taken the cases just in the order they have come under my notice, only rejecting those where the bleeding from the urinary passages occurred under circumstances which cannot be said to have anything but a very limited significance. I refer to such cases as those where blood is mixed with the urine as a consequence of the use of catheters and other instruments, and in connection with gonorrhœa, and the use, or rather abuse of injections; these I have excluded from my list.

In the hundred cases I have thus taken I will put in the order of their frequency the causes which in my belief have produced this symptom or are responsible directly for its continuance:—

Kidney stones	30
The hypertrophied prostate of elderly males	20
Bladder stones	14
Tumours of bladder and prostate—mostly malignant	13
Urinary tuberculosis	6
Stricture of the urethra	5
Cystitis not due to a senile prostate ...	3
The passing of crystals from the kidney downwards...	3
Traumatisms or their effects	2
Filaria sanguinis hominis	1
Of very doubtful origin	3
	100

I am not at all sure whether, in this country at all

events, the order of frequency which I believe occurred
in my practice does not represent tolerably accurately
the general experience in reference to this one symptom.
I say this country, because local conditions may have a
largely determining influence. For instance, if statistics
on a large scale were taken in certain parts of the East,
as in Egypt, it would be found that the hæmaturia due
to a parasite largely predominates. In certain districts
of England, as well as in the practice of some medical men,
the hæmaturia due to stone and gravel might be even
much in excess of that which I have put it at, but taken
as a whole the proportions I have given will, I think, be
found fairly representative.

Hæmaturia may be regarded as a tolerably constant
symptom of stone in the kidney, and is, I believe, more
to be relied on for diagnosis than any other. It is not,
however, invariably present. In the case of a man from
whom I removed two stones from the kidney a few
months ago with success, this symptom was absent, the
indication for the operation being the acute paroxysms
of lumbar and testicular pain from which he suffered.
The persistence of hæmaturia associated with symptoms
of renal colic is sufficient to warrant the exploration of
the suspected organ by an incision in the loins. I have
practised the lumbar incision as elsewhere described (a)
and have every reason to be satisfied with it.

In the thirty cases of hæmaturia due to impacted
renal stone which I have tabulated, there were several
where the escape of the calculus followed the use of boracite as described in the previous lecture, and with this
the cessation of the hæmorrhage.

(a) Notes on the Surgery of the Kidney, *Liverpool Medico-Chirur-*
gical Journal, Jan. 1889.

It may not be out of place to point out here that the descent of a calculus from the kidney may be favoured sometimes by the distension of the bladder with water and its regurgitation along the ureters. I have endeavoured to show in a paper on the subject (a) how this may be done, and how very little it sometimes takes to favour the dislodgment of a stone from the kidney which may have been a source of hæmorrhage and pain for some time.

Stone in the bladder was attended with bleeding in fourteen instances. Hæmaturia is not as a rule a constant symptom of this affection, though the circumstances under which it occurs are generally very significant. When you hear of a person with an irritable bladder who cannot, for instance, take horse exercise, ride on an omnibus, or walk along a rough road without seeing blood in his urine, you may take it as a good hint that he may have a stone in the bladder. At all events, in the absence of other indications it is generally worth acting upon and suggesting that a sound should be passed, and the bladder explored. Then an acute attack of hæmaturia, followed by irritability of the bladder, may sometimes indicate that though a stone has passed down from the kidney it has not escaped by the urethra. This happened in two instances out of the fourteen I have tabulated. In both cases a single crush with the lithotrite effected the immediate discharge of the fragments with no further inconvenience to the patient.

If, after patients have suffered from undoubted signs of renal colic, and no stone is known to have been passed, they could be induced to have the bladder explored, and, if necessary, a lithotrite introduced, the mortality after stone operations would be practically nil.

(a) *The Lancet*, March 10th, 1888.

In three instances out of my hundred, the passing of crystals, chiefly uric acid, was apparently responsible for the appearance of blood in the urine. In none of these cases was the amount of blood considerable. Some persons seem never to be well unless they are passing large quantities of uric acid, and everything which tends to check the excretion appears to add to their discomfort. I believe the hæmorrhage that attends these discharges of uric acid is connected with the precise nature of the cystalline form, as a change in the shape of the crystal is often followed by a cessation of this symptom. It is under these circumstances, and with this object that the waters of the Vosges, such as Contrexeville and Vittel often prove of so much service. The youngest patient of the three coming under this category was a female child, a few months old, whose urine contained a large number of uric crystals which seemed to be the cause of the bleeding. The patient was fed on equal quantities of milk, barley, and Contrexeville water, under which the symptoms disappeared. *Apropos* of hæmaturia caused by crystals, I would refer to a correspondence (*a*) relating to the presence of blood in the urine after eating cooked rhubarb. I have known this happen on several occasions, especially where the drinking water is hard, as pointed out by Dr. O'Neill, crystals of oxalate of lime being found in the urine in abundance. I have also seen the most profuse hæmaturia occur after an excessive indulgence in asparagus.

Senile enlargement of the prostate appeared to be responsible for twenty cases of hæmaturia out of the number. In the majority of instances the amount of bleeding was not large, but it would recur from time to time on very slight causes, such as cold and fatigue.

(*a*) *The Lancet*, July 5th, 1890.

There are a good many persons with large prostates who get temporary attacks of bleeding much on the same principle as others do who suffer from piles. Dr. Frank reminded me the other day that the term "prostatic pile" was a condition recognised by some German writers on this subject. I fully appreciate the force of the expression. When the bladder is capable of emptying itself tolerably well this symptom is merely a temporary one and usually disappears with some restrictions in diet and a little active purgation. There is, however, one condition of the senile bladder which adds considerably to the trouble connected with this kind of bleeding. I refer to those instances where it occurs with a large prostate and an atonic, or almost completely atonic, bladder. The great safeguard against prostatic hæmorrhage is the power of the bladder to exercise pressure. In two instances not only had I to empty the bladder of blood, but to keep it empty by pressure upon it, and the retention of a catheter until the tendency to bleed had ceased, just as is done with the flaccid uterus. In both the instances I refer to this was successfully accomplished, and the patients recovered, though the loss of blood was considerable. It is not the least use depending upon hæmostatics in cases such as these, the mechanical reason why the bleeding will not cease must be recognised and acted upon, or the patients will flood to death with their bladders distended with blood up to the umbilicus, as I have seen.

For the purpose of emptying the bladder under such circumstances there is nothing like one of the catheters and aspirators which are employed in evacuating the fragments of stone after lithotrity. Unless this is effectually done, there is very little chance of restraining the hæmorrhage.

53

Malignant tumours of the bladder or the prostate accounted for thirteen cases of hæmaturia. If the disease is in the prostate, you can generally make it out with the finger in the rectum ; if this is not the case, then you may have this very significant fact, that though from its nature the blood evidently comes from the interior of the bladder, you have no senile prostate to account for it ; the patient may be, and often is, considerably under the hypertrophic age. Then there is another helpful point in connection with the diagnosis of malignant tumours of the bladder, they are as a rule more comfortable when bleeding moderately than when the urine is absolutely clear. This is a point I have often noticed in cases which have been verified, either by operation or post-mortem examination. It is also curious to notice how slow some malignant growths of the bladder and pros-tate proceed : there may be repeated attacks of hæmor-rhage, but I have known such cases as epithelioma of the bladder go on for four and six years, and really occasion but little inconvenience until the bleeding becomes pro-fuse and unrestrainable, and the question will then natur-ally arise, What is to be done when the hæmorrhage from such a growth becomes excessive? My belief is that when this stage arrives, in the majority of cases the best thing to be done is to open the bladder above the pubes. I can hardly recall an instance when, by this ex-pedient, a patient did not obtain a temporary respite from bleeding, and the consequences that follow it. Even if the wound never closes again the patient can be kept in a far greater state of comfort than when he only has his urethra to depend upon. In the female the ready way in which the growth can be got at and removed is a strong reason for early exploration, and, if necessary, removal.

Hæmaturia is almost as constant an indication that tubercle has invaded the kidneys or ureters as hæmoptysis signifies that the lungs are tubercular. Urinary tuberculosis may be either ascending or descending. It is always a good plan in tubercular subjects or in persons who are suffering from ill-defined slight attacks of hæmaturia, and possessing tubercular histories, not to neglect to make a careful examination of the testes, the prostate and the vesiculæ seminales, with the finger. The detection of tubercular deposit in any of these positions, under these circumstances, will often throw the necessary light upon the hæmaturia, The shotty feel of the prostate when impregnated with the early form of this deposit is important to know how to recognise. In the breaking-down stages of urinary tubercle, the presence of pus as well as blood, not to say anything of the bacillus, adds considerably to the ease of making a diagnosis.

Stricture of the urethra is accredited as the cause of five cases of hæmaturia. It is not as a rule the stricture that bleeds, but the bladder from the atony that ensues by reason of the viscera being very imperfectly emptied. In two instances the strictures were extremely tight, and it seemed impossible how the cause of the bleeding could have been overlooked. In every instance where the stricture had been sufficiently dilated and the bladder had regained its tone, the blood disappeared from the urine.

It is curious to notice how some persons with really tight strictures, by the use of greater expulsion force mitigate considerably the inconvenience the contraction would otherwise cause. It is not always easy to remember this in practice. There is a patient attending the out-patient department to whose symptoms I have often directed attention. He came here complaining of loss of power over the bladder and lower extremities, from the loins downwards. There was diminished sensation and the reflexes were unnatural, and it was clear from his gait that he was suffering from an early stage of paraplegia. On examinining his urethra I found that he had a moderately tight stricture in the membranous portion, and as there was no other cause for the condition of his limbs I regarded it as one of reflected paralysis, and commenced to dilate his stricture. Under this treatment, it was observed that movement and sensation in the affected parts gradually returned and in three months all symptoms of paralysis had disappeared, the patient being able to walk several miles without fatigue. I have met with two or three instances of this kind. I am induced to refer to them here as illustrating how a stricture may not only escape notice, but produce remote symptoms which, at first sight, would hardly suggest such a cause. In like manner a persistent hæmaturia may be dependent upon a contraction in the urethra and continue until the latter is removed.

Wounds involving the kidney may in some remote way or other cause blood to find its way into the urine and become mixed with it. Let me mention two examples:—In May, 1888, I saw a patient in consultation with Dr. Glynn who was suffering from purulent urine. Six weeks before I saw him he had a very free attack of hæmaturia which lasted for some days. He had no symptoms of renal colic, but complained of a dull aching pain in the region of the left kidney. His history was as follows :— Five months previously when at sea he was shot by his steward in two places, one ball passing through his thigh, and the other entering his left side two inches in front of the posterior superior spinous process of the ilium. He was leaning over his assailant who was on the ground at the time the bullets were fired. He was confined to bed for some time and has since suffered from urinary irritation and purulent urine. There appeared to be an increased area of dulness below the left kidney, and the part was sensitive to the touch. The bladder was sounded, but I could detect nothing abnormal. I have no doubt the bullet is embedded in the left kidney, from which place I proposed to remove it, but as the man was returning home to America, and was in fairly good health, he declined to submit to any operation at present. The other case of traumatic hæmaturia followed the escape of a pin by the urethra which had been swallowed by a gentleman, 56 years of age, four months before it was voided.

Three cases of hæmaturia were associated with cystitis which had occurred in males independent of a large prostate or gonorrhœa. In one the cystitis followed an injury to the abdomen where probably the bleeding was due to some slight laceration. In the other two instances it seemed to be connected with the violence of the inflammation, one of the patients being an extremely gouty subject with highly acid urine.

My list contains one illustration of bleeding due to the filaria sanguinis hominis. The presence of this parasite in the urinary system explains cases which were formerly described under the name of chylous urine. In referring to this parasite, Sir William Roberts (a) says: " Its local effects are supposed to depend on the formation of aggregations of filaria which block up the capillaries and cause by their active movements irritation and rupture of the blood channels and lymphatics, and thus lead to the appearance of chyle and blood in the urine." In the case to which I am now referring, the urine was at one

(a) " On Urinary and Renal Diseases." 4th edition.

time quite chylous in appearance whilst at others it was
deeply coloured with blood and contained large clots.
The presence of the parasite was in this instance detected
both in the blood and urine by Sir William Roberts and
Dr. Stephen Mackenzie, who were kind enough to see
the patient. The administration of iodide of potassium
in 20 grain doses, as first suggested by Dr. Harley,
seemed to be followed by good results.

The list closes with three cases where the origin of
the bleeding seemed so obscure that I did not even like
to venture an opinion, which could not be supported
either by reason or experience, as to its probable origin.
In many of the instances where there was a doubt as to
the nature of the bleeding and in others where further
confirmation seemed desirable, the electric cystoscope
was used. As an adjunct to other means for diagnosis
I can testify to the value of this instrument. On several
occasions I have gladly availed myself of the kind assistance of my colleague, Mr. Hurry Fenwick, whose recent
work on this subject has contributed so much towards
establishing the utility of electricity as an illuminant
for surgical purposes.

LECTURE V.

THE EARLY DETECTION AND TREATMENT OF STONE IN THE BLADDER.

THE risk attending the removal of a stone from the bladder may be roughly estimated by the size the stone has been allowed to attain. Hence I have selected as a useful subject for our consideration to-day certain points connected with the early diagnosis and treatment of this affection.

I need hardly remind you that when a stone assumes such dimensions, or is of a shape where it cannot be expelled along the urethra by the natural efforts of the bladder, its removal must be effected either by a crushing or a cutting operation. Up to within a comparatively recent date the various operations included under the term lithotomy were almost universally employed, the crushing operation being confined to the least dangerous forms of the disorder, and in a sense restricted to the practice of a comparatively few. During recent years all this has been greatly changed, lithotrity is now the rule at all events in adults, the cutting operation being reserved for exceptional conditions either of the stone or the parts involved. When we consider the small risk that now attends the operation of lithotrity it is impossible not to feel in every case where a stone has to be removed from the bladder by the use of the knife, that there was a time in its history when such a proceeding would have been quite unnecessary. In the present day the rejection of lithotrity in a given case implies one, at least, of two things, (1) either that the stone has been allowed to grow too big to admit of its removal in this way, or that (2) by its long continued presence the stone has so altered the structural arrangement of the parts in its immediate vicinity as to render the ultimate success of the crushing operation highly improbable.

Before proceeding to notice some mechanical details connected with the detection of stone, I would remind you for reasons that will be obvious that the primary formation of stone in adults and children is very different. In the former I believe, as a rule, vesical stones are extensions on appreciable nuclei which have descended from the kidneys, whilst in the latter they are from the commencement aggregations formed in the bladder. This is an important distinction which often works out in the following way. In children it is almost impossible to form an approximate idea as to when a stone began to form in the bladder, whereas in adults the process can often be distinctly connected with a marked attack of renal colic, hæmaturia, or nephralgia. This is especially noticeable in those cases were urates or oxalates form the nucleus or centre of the stone, as when the calculus is phosphatic throughout, it not infrequently happens that there is present some special reason why the bladder should be entirely responsible. In the case of many stones I have thus not only been able to fix the probable date when they entered the bladder, but to form an estimate of the rate at which they have grown.

F

Hence, as I have said in a previous lecture, persons who pass stones from the kidneys would do well if there is any doubt, to take steps for ascertaining that the stone has left the bladder and is not remaining there to grow and to make its subsequent removal a proceeding in which an element of risk is introduced.

If I were asked to briefly enumerate the indications for sounding the adult male bladder, I should put them down in the following order : for continuing irritablity of the bladder unexplainable or insufficiently so, by the presence of a large prostate : for the recurrence of blood in the urine, particularly after brisk or jolting exercise, and for the frequent intervention of bladder spasm, either with or without the habitual use of the catheter, where the prostate is known to be enlarged. These conditions will be found to indicate more frequently than any other the presence of a stone in the bladder. Pain or irritation felt at the end of the penis is of more value as a symptom of stone in children than in adults, as with the latter a large prostate rendered sensitive by the condition of the urine passing over it is quite sufficient to account for this symptom. For instance, nervous persons in the habit of passing highly phosphatic urine usually have tender prostates, and often complain of pain at the end of the penis. I have known such individuals to be frequently sounded for stone or tumour on that account.

For sounding purposes nearly every surgeon has his favourite implement which generally bears his own name. An inspection of a variety of sounds would seem to indicate that the hand has more to do wlth the process than the exact shape and make of the instrument. Sounds should not be so large as to fit the urethra too tightly, otherwise much of the delicacy of the touch will be lost.

Where the prostate is very large the surgeon should be provided with a sound two or three inches longer than those usually made, otherwise he may fail to easily reach all parts of the bladder, unless the penis is inconveniently telescoped. When the bladder is much distorted in shape you can feel a stone with a soft bougie when a steel sound cannot be made, without force, to touch it. For such a purpose I have sometimes found a flexible instrument with a metal tip extremely useful. As Dr. Freyer has pointed out, the aspirator and catheter employed for the removal of fragments of stone after lithotrity make a capital sound by causing stones, often of small size, which might otherwise escape notice to rattle against the eye of the instrument. I have employed this device for the removal of certain kinds of foreign bodies from the bladder, and thus avoided resorting to less delicate forms of searching and extracting instruments. Here is a fine pig's bristle, slightly coated with phosphates, which was detected and removed by the apparatus just referred to. In this instance the patient had first passed up into his bladder a piece of bacon rind with the view, I presume, of lubricating the urethra, the bristle being subsequently introduced. Symptoms of bladder irritation coming on, I sounded with an evacuating catheter and rubber syringe such as I am now showing you, with the result mentioned. The case is fully reported elsewhere (a) with remarks on this subject.

It must not be forgotten that, however delicate your touch and however well devised your sounds may be, there are some kinds of stone which you do not feel until you have them in the jaws of the lithotrite. This has happened to me twice recently; so far as the sounding went, and with a strong suspicion in my mind that a stone was present

(a) *The Lancet*, October 29th, 1887.

in each instance, I could not **detect** one until I had passed a lithotrite. and had actually got the calculus within the grasp of the instrument. **Here** is the specimen from one of these cases. Some phosphatic stones are covered **over** with **a** thick slimy sort **of** tenacious mucus which effectually muffles the sensation and sound that **contact** between **a steel** body and a stone would otherwise **give.** **Large** stones which are located behind and above the **pubes are** sometimes missed by the curved sound passing **directly** under them. When you are sounding, a lithotrite should always be at hand, not merely for measuring the **stone** **when detected, but** for gently feeling about with. **As a rule it is** better ,**to** **sound a** person under **an** anæsthetic; **in this** **way** pain is entirely avoided, and the **examination is rendered far more** searching **and** complete.

Apart, however, **from the question of size, it** is most important **that** a **stone** **should be** detected before it leads to certain structural **changes** in the bladder itself. Five years ago (a) I described some forms **of** calculi in the bladder under the name of fixed **or station-** **ary stones,** implying that without being encysted **they** **occupied** **certain** positions, most commonly above **the** **enlarged prostate,** from which they seldom moved except **by accident.** In connection with this **point one** cannot **go over a fairly** long list of lithotrity cases without **here** **and** there recognising one **where** bladder symptoms only became urgent and continuous **after** the stone had been completely **and** properly **crushed** and evacuated. Let me illustrate what I mean. **A** patient about **sixty** years of age, consulted **me** some years ago for occasional hæmaturia. Behind **an** hypertrophied prostate **I** found a large **urate** stone **which** I subsequently removed by lithotrity.

(a) "Annals of Surgery," June, **1885**

After some weeks of complete relief, he again began to complain of irritability of the bladder, and in addition the discharge of ammoniacal urine indicating that which he had never previously suffered from, namely, chronic cystitis. This continued for some time, when on sounding him again I found the place of the original stone occupied by a considerable mass of triple phosphates which I also removed by crushing. The same thing occurred about a year afterwards. The first stone that formed no doubt represented a constitutional tendency, whilst the latter ones seemed to me to be connected with certain changes in the shape and structure of the bladder wall, which had been brought about by the long continued presence of the one that had preceded them. The effects produced on the bladder itself by the retention of a stone for some considerable time with special reference to the recurrence of calculus, is a subject worthy of careful consideration as insufficient attention has hitherto been paid to it.

That the earlier detection of stone which now prevails, coupled with Bigelow's method of operating, has contributed largely to the reduced mortality connected with the affection, there can I think be no doubt. Any one who is at all acquainted with the subject will recognise as a fact that the risks attendant upon the retention of a small stone in the bladder, are really far greater than those which its removal by lithotrity now entail. For, apart from its influence on the structure of the bladder and the irritation it creates, the presence of a calculus may not only prove to be a constant source of bleeding, but may actually lead to the rupture of the bladder, as I have shown in a previous lecture.

Further, as bearing upon the importance of dealing with stones whilst they are within the reach

of lithotrity as now practised, we may point to
our greatly lessened period of convalescence as compared with that which prevailed before Bigelow's
discovery. Looking back at my notes of cases which
occurred before and after this period, and comparing
them stone for stone, the period of convalescence has
diminished by less than one half. It is not at all un-
common to see stones which formerly would have occupied two or three operations, or sittings as they were
called, extending at intervals over a month or so, now
disposed of at one crushing, recovery taking place within
a week or even less. This not only means a considerable
diminution in the amount of direct risk which repeated
crushings necessarily entail, but in those connected with
whatever tends to protract a convalescence. I do not
think I am wrong in stating that Bigelow's method has
reduced the period of recovery after lithotrity, taking all
sizes of stone together, by something like one half the
time. No better illustration of this point could be taken
than the last case I operated upon at this hospital, at which
several of you were present. The patient, a male, 50 years
of age, was operated upon on July 2nd, and was discharged well on the sixth day afterwards. The stone
was a urate one coated with phosphates, and had a
diameter of an inch and a quarter. Though the operation occupied nearly half an hour the water in the wash
bottle used for removing the fragments, was noticed to
be hardly discoloured with blood, and the day following
the operation the urine was perfectly clear and normal in
appearance. There are a few details I would lay stress upon
which may be of service to those who are commencing to
practice lithotrity.

1. To suspend all manipulations with the sound or
lithotrite, or in washing out the fragments when the

bladder is on the move. Patients even when profoundly under an anæsthetic with the reflexes in abeyance, will sometimes commence to force and squeeze with their bladders in a most remarkable manner. The pressure of this may be roughly measured by the distance fluid is projected from the catheter even by some bladders which were supposed to be atonied or paralysed. All movements on the part of the operator should be at once suspended till this is over. I have known three instances at least where rupture of the bladder occurred from disregard of this precaution, two where the lithotrite was being used, and one where the washbottle was in action. It is very easy to rupture a bladder when it is on the strain, whereas such an accident is most unlikely to happen where the organ is quiescent.

2. Conduct all your manipulations so as not, if possible, to cause bleeding. Some operators will hardly discolour the water in the wash bottle even where the manipulations are prolonged.

3. Before the operation is completed remove all clots from the bladder and wash out with warm water until there is evidence that any bleeding which may have been caused by the manipulations has practically ceased.

4. Ascertain before the patient recovers consciousness what kind and size of catheter passes easiest. On the use of this instrument in the after management of the patient much depends. If urine mixed with blood is retained and allowed to undergo putrefaction some degree of septic fever is sure to occur. As a rule, after each time the catheter is used two or three ounces of warm boracic lotion in the proportion of a drachm to a pint should be thrown into the bladder by means of a rubber bottle and allowed to escape. This should be continued

until the urine is voided, either with or without the use of the catheter, of acid reaction and free from blood staining.

Passing from by far the larger class of cases which are now treated by lithotrity, and from urging the importance of endeavouring, by an early diagnosis, to bring all cases within this category, I will proceed to notice those instances where lithotrity seems to be indicated.

Stones in male children; very large stones in male adults, where the weight may be estimated at several ounces, and some smaller sized recurrent stones complicated with sacculated, or bladders requiring drainage, but otherwise quite within the limits of lithotrity, may be generally stated as better adapted for lithotomy in some form. So far as male children are concerned, some surgeons will probably join issue with me. The large experience of Drs. Keegan and Freyer and other operators in India more particularly, as well as the smaller experience of some surgeons in this country and elsewhere, would seem to indicate that a skilled lithotritist may obtain most excellent results in even the youngest subjects from crushing. A short time ago my colleague Mr. Heycock, crushed a hard stone weighing nearly an ounce, and the boy was well and about within a week. But what about practitioners who are not skilled in lithotrity and yet have large experience generally in operative surgery? I feel under these circumstances that lithotrity as a rule will prove to be as it always has been, the wiser selection.

In choosing the kind of lithotomy I must express my preference both in boys and adults for the lateral method, unless I want to remove a stone which from its size is not likely to come out very easily through the perineum, or if I wish to look into the bladder with the naked eye for the purpose of searching for a sacculus in which a stone may be lodged, and at the same time dealing with it. To the latter category, where the suprapubic operation is indicated, Mr. Mayo Robson (a) has added illustrations where not only has a stone been thus removed from the bladder, but a portion of the large prostate extirpated.

When prolonged drainage of the bladder is necessary in certain cases of chronic cystitis complicated with stone, or in some instances of recurring phosphatic stone, where the posterior wall of the bladder is much dependent, and a division of the floor of the prostate is required, I feel sure that the perineal route is to be preferred. I have seen successful instances of this where drainage had previously been practised and maintained for some time by a suprapubic opening.

In comparing the suprapubic and lateral operations there can be no doubt that the former proceeding will recommend itself to many practitioners on the ground that it is the simpler of the two, and if for this reason it is the more efficiently executed, no better ground can be advanced for its selection, provided that the same end can be obtained. When I have chosen the suprapubic operation no attempt has been made to secure the primary healing of the wound, and so far as sutures are concerned, they have only been used for approximating a limited portion of the superficial wound. I then put in a large rubber drainage tube, which is passed through a covering made by a wood wool pad. The tube is retained for two or three days and then removed, the patient being kept absolutely dry by the constant use of the wood wool pads, which have been found to be the simplest and most

(a) *British Medical Journal*, March 9th, 1889.

G

efficient mode of dressing. By using any dressing which does not absorb, infiltration of urine into the cellular tissue in front of the neck of the bladder may readily occur and retard the progress of the case.

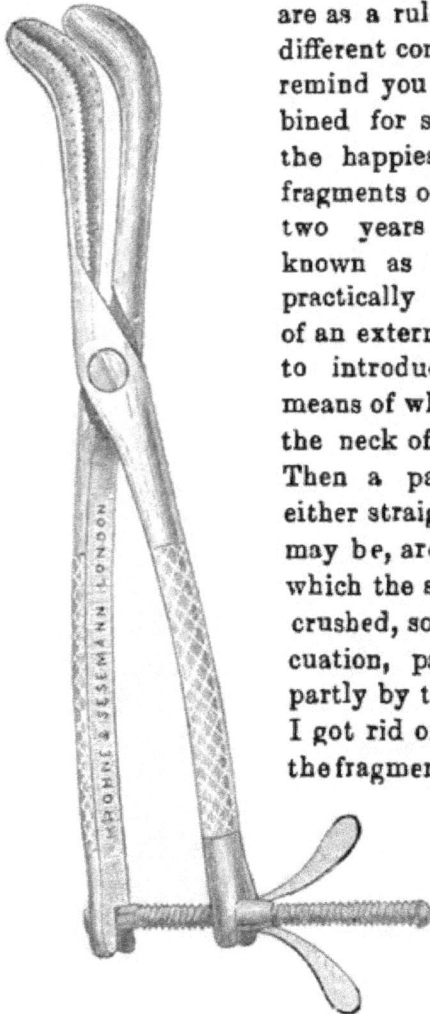

FIG. 4.

Though lithotomy and lithotrity are as a rule employed to meet quite different conditions, I would like to remind you that they may be combined for some special reason with the happiest results. Here are the fragments of a large stone I removed two years ago by the operation known as perineal lithotrity. It practically consists in the making of an external urethrotomy sufficient to introduce the index finger, by means of which the parts constituting the neck of the bladder are dilated. Then a pair of crushing forceps, either straight or curved, as the case may be, are passed in, by means of which the stone or stones are rapidly crushed, so as to admit of their evacuation, partly by the forceps and partly by the aspirator. In this way I got rid of a large phosphatic stone, the fragments of which weighed nearly three ounces in something under five minutes, a process which I do not think could have been as safely and as certainly accomplished under half an

hour by the ordinary lithotrite and evacuator. The
patient, a young man of eighteen years of age, was up and
well with the wound soundly healed on the eleventh day.
The crushing and extracting forceps which have been
made for me by Messrs. Krohne and Sesemann are shown
in the figure (Fig. 4). •

There are cases of stone, chiefly recurring ones, where
not only is a digital examination of the bladder necessary,
but where after the removal of the stone the bladder may
with much advantage be submitted to a process of drainage with the view of preventing further relapse. In such
instances I have availed myself of perineal lithotrity,
the nearer approach to the bladder which the puncture
permits of proving an important feature in promoting the
success that has followed the adoption of this combination
with practically no additional risk.

LECTURE VI.

SOME MISCELLANEOUS POINTS.

In the first place I should like to say a few words in reference to the pathology of gleet, inasmuch as this extremely common affection is not only difficult to cure, but frequently proves the starting point for much more serious conditions. Upon the pathology of gleet considerable light has been recently thrown by Oberlaender, by means of the electric endoscope, and various effects of persistent inflammation on the mucous tract have been pointed out. In the earlier stages of gleet, observation with the endoscope shows that the process practically consists in more or less acute epithelial desquamation, at points in the deep urethra where the affection is usually situated. In this way the slight amount of discharge which may be said to constitute the disease is supplied. Though there is generally some tumefaction about the affected spots, no stricture can then be said to exist, and treatment is still capable of bringing about a complete restoration of the parts without the subsequent introduction of any contractile tissue. And this brings me to consider the next stage in the process which eventually leads to the formation of an organised stricture. The function exercised by the epithelial coat of the urethra, and the consequences that may ensue when it becomes deranged

by persistent inflammation, formed the subject for some
observations in my Lettsomian Lectures before the
Medical Society of London, in 1888. The view that I
then advocated was, that by chronic inflammation, such
as we see in gleet, the epithelial coat becomes so
damaged at one or more spots as to lose its power of
rendering the urethra urine-tight, and that a slow process of exudation through the mucous coat ensues. As
a consequence of this, and to prevent urine soaking
further into the tissues, inflammatory lymph-exudation
in the sub-mucous area is excited, and barriers of lymph
which eventually become organised are thrown out oppo-
site the places where the leakages have taken place. The
contraction of these sub-mucous lymph barriers, I urged,
completed the process of stricture formation. This view
as to the pathology of stricture is supported by the
following considerations:—(1) That though the mucous
membrane is the tissue chiefly concerned in the primary
inflammation, it is as a rule only secondarily implicated
in the stricture forming process. In many instances it
will be found after death that the dimensions of the
mucous membrane are not permanently altered, and
that it is possible to split a stricture without necessarily
rupturing the lining membrane of the canal; and (2) the
great variety in form and shape in which stricture tissue
is deposited in relation with the urethra. I have known
as many as a dozen irregular points of contraction in the
urethra of a man, which were believed to be formed in
the way I have described, consequent upon a very
chronic gleet.

Though the views I have thus given expression to
recognise the existence of varying degrees of tumefaction
or swelling within the urethral walls at one or more spots
in the initial stage of the stricture forming process, I

cannot admit the propriety of proceeding to the performance of internal urethrotomy for their removal, even though, as I have no doubt, they are capable of being demonstrated as obstacles to micturition by such instruments of precision as Otis' urethrameter. My experience, derived from the observation of cases of this kind which had been submitted by various surgeons, to internal section, is certainly not favourable to this method, however well adapted it may be to some advanced forms of cicatricial stricture. Some of the worst cases of stricture in the deep urethra that have come under my notice have been those where an internal urethrotomy had been performed during the early or gleety stage of the affection. In the pendulous urethra such ill effects following the operation have not been so frequently noticed.

In urging these opinions as explaining the mode in which stricture tissue is commonly deposited, I must not be regarded as failing to recognise that large class of strictures which are the result of wounds in the urethra. Here the healing process goes on with the wound exposed to the irritating influence of contact with pent-up urine, and the result is the formation of a cicatrix of the densest and most contractile description. That a dense contractile stricture is by no means a necessary consequence of a torn or lacerated urethra, I have demonstrated by a series of cases where such injuries had been treated by perineal section and urine drainage, without leaving behind any trace of the original injury. Many of these cases have been watched for periods of time extending over two years.

In giving expression to these views as to the pathology of gleet and the healing of wounds of the urethra, it is for the purpose of emphasising the importance of thorough irrigation in connection with the prevention

and treatment of urethral stricture. I have long felt
that the difficulty in this class of affections lies, not in
the want of suitable medications, but in our imperfect
means of using them. It is astonishing what perfect
douching with cold water or some mild astringent will do
in some cases where individuals are driven into
the depths of despair by some small amount of
urethral discharge, which frequently is no more
hurtful than are the ,slight chronic excretions which
proceed from other parts of the body that might
be mentioned. The first apparatus I described for this
purpose consisted of a Higginson's syringe, fitted with a
small rubber catheter, by which a running stream of
water could be maintained from any point in the deep
urethra, outwards. (a) More recently I have adopted
for the same purpose as giving a more equable stream, a
simple syphon apparatus, which was first made by Mr.
Reynolds, of Liverpool, on the principle of Thudichum's
nasal douche. It is principally in the treatment of gleet
that I have found this instrument of most value. As a
rule, the reservoir is filled with about six ounces of a
solution of sulpho-carbolate of zinc, two grains to the
ounce, with which the patient irrigates his urethra for
some minutes once or twice a day. By compressing the
meatus around the pipe of the irrigator the outflow is
temporarily stopped, and the fluid is forced by hydrostatic
pressure into the various lacunæ of the urethra, which,
in chronic cases particularly, are more or less centres of
infection. This thorough mode of irrigation is almost
impossible to be obtained with an ordinary glass syringe,
as the latter is often only worked at a great disadvantage,
as I have had the means of demonstrating in a very

(a) "On the Prevention of Stricture and Prostatic Obstruction.
Churchill, 1881.

simple manner. Alum, zinc, borax, acetate of lead, or Condy's fluid, may in some instances be substituted for the sulpho-carbolate of zinc, but the latter has proved to be the most reliable. Now this simple, but not I believe sufficiently well-known means of douching the urethra, is capable of extension to other conditions than that mentioned, where it is necessary to bring the canal under direct medication. There are other discharges—prostatic, seminal, and mucous—which are often maintained by an unhealthy state of the urethra; for such of these as require local treatment, I have found it in the irrigator, rather than in the employment of those powerful and dangerous cauterisations, which at one time were advocated by Lallemand and others. In cases of orchitis, where the infection has proceeded from the urethra, as in gonorrhœa, I have found deep urethral irrigation with a solution of perchloride of mercury (1 in 10,000), most effective in shortening and moderating the attack.

Passing from this reference to the use of the douche in various urethral discharges, I would allude to it further as an important adjunct in the treatment of stricture by dilatation. After every dilatation, when the calibre of the stricture will permit the entrance of the fine point of the irrigator, I have the urethra well douched with warm water, commencing from the point immediately behind the obstruction. This is continued for a few minutes, and has a most beneficial effect, not only in thoroughly cleansing the canal, but in assisting, by hydrostatic pressure on the contracted parts, the process of dilatation. Strictures which have been treated in this way, have not only yielded to dilatation more rapidly, but what is of still more importance, they have, as might be expected, shown less tendency to contract, and have been more amenable to the subsequent use of the bougie. It is my habit to instruct patients to employ it themselves, as a necessary part of the treatment by dilatation.

A word may be said in reference to the use of the douche in connection with the treatment of the less complicated forms of urethral fistulæ. There is hardly anything which causes more annoyance to a patient otherwise sound, than the constant dribbling of small quantities of urine along a sinus opening in the perineum, or by the side of the scrotum, which has resulted either from a peri-urethral abscess, or from some simple surgical procedure connected with the urethra. The two great obstacles to sound healing from the bottom of the fistula are (1) the persistence of some degree of stricture or contraction of the urethra in front of the communication with the false passage, and (2) an unclean, and therefore septic condition of the internal aperture caused by the lodgment of urine within the urethra after each act of micturition. When the stricture has been sufficiently dilated I have frequently succeeded in getting these sinuses to heal quickly and soundly, by making the patient irrigate his urethra with some suitable antiseptic three or four times in the course of the twenty-four hours.

And now I will draw attention to a few instruments I brought under notice some years ago, (a) which I believe are by far the best and safest bougies we can employ in most cases of tight stricture. They are popularly known as "whips," being the name given to them by my friend Mr. Lund, in his Hunterian Lectures. They are made in different sizes, in lengths of about twenty inches, commencing by a filiform point, and gradually increasing towards the other extremity. In speaking of the value of these instruments, in the treatment of tight strictures

(a) The *Lancet*, February 3rd, 1883.

especially, I am sure that a trial by those unacquainted with them will ensure their adoption in that class of strictures where the difficulties of instrumentation are the greatest. In my own hands they have had an extended trial, and when I have demonstrated their employment to others, their adoption has generally followed. It is no small recommendation to say of an instrument of this kind, that it rarely fails to pass a stricture, so long as it is structurally pervious, however fine the passage may be ; that dilatation necessarily follows upon this without risking the re-introduction of a bougie on a larger scale ; and that these advantages can be secured without the liability to laceration and hæmorrhage which often accompany the use of other instruments. This is important to remember, as every degree of laceration the urethra receives adds to that kind of tissue of which an organic tissue consists. Further, by reason of the tapering form of this bougie, every introduction of it not only stretches the stricture, and increases its calibre by several sizes of the French gauge, but it tends to make the entrance to the stricture from the front funnelled shape, and thus adds to the ease with which subsequent introductions are effected, either by the patient or the practitioner. Patients find these instruments, after they have been taught their use by the surgeon, far more easy to introduce than any other, and as they cause little or no friction in making their way through the obstruction, the process of dilatation proceeds more rapidly than is the case when more abrupt bougies are used. It is better to soften new whips, by immersing them for a few seconds in warm water, so that they may curl up in the bladder without the patient being conscious of it, as is the case with the pilot bougies which are now so generally appended to many

urethral instruments. By the use of this form of
instrument, it is quite practicable, by the introduction
of two or three larger ones, to bring up at once a very
contracted stricture, until it will admit with ease a good-
sized bougie or catheter. I believe that we cannot proceed too rapidly with this process. To relieve a patient
with retention of urine by means of a fine catheter,
perhaps tied in with the chance of it slipping out, is to
leave him in a very sorry plight. A person who is going
about dribbling, with a stricture of small calibre, is in
constant danger of his life, as there is no knowing when
sudden retention, perineal abscess, or extravasation of
urine may supervene. Some melancholy illustrations of
this must occur to every one who has seen much of
urethral stricture. Mr. Robson (b) has recently illustrated the value of immediate dilatation with bougies,
and in the commencement of this process I have found
the whip invaluable.

In reference to the selection of instruments generally,
I much prefer the flexible ones ; they are readily introduced, and rarely cause hæmorrhage. Those made of
silk web have some advantage over the rubber catheter,
as they can be used with a little more firmness. It is
curious to observe the difference in the curves of the
prostatic urethra ; hence in one case a *coudé* slips in
easiest, in another an olive-pointed instrument, whilst in
a third an ordinary round-headed catheter glides in
without trouble. It is important that the delivery power
of a catheter should be relative to the normal size of the
urethra, and not greatly in excess. What can be worse
for a person who has been accustomed for some years to
ooze or dribble his urine, than to have it suddenly drawn
off *pleno rivo* in about half the usual time the normal act

(b) *Provincial Medical Journal*, vol. viii, 1880.

occupies. I have often found the too rapid emptying of the bladder in the regular employment of catheterism an important factor in producing loss of natural expulsive power, and in retarding its return when it has been temporarily suspended. Nature will not provide muscle when it is not wanted. Here, then, we have at least two important reasons why the selection of catheters should not be handed over to the instrument maker alone, as a patient may very innocently do himself harm by means of a catheter unsuited to his case.

Whatever means or form of instrument, whether flexible or not, our individual experience prompts us to use in endeavouring to overcome the obstacles that some strictures present, should they prove ineffectual, I believe supra-pubic aspiration is the best and safest expedient of a temporary nature that can be employed. By means of this we can always depend on being able to empty a distended bladder with a very small amount of risk, and less, I believe, than what may in some instances attend difficult catheterism.

In reference to aspiration in connection with some urinary disorders, I believe that it occasionally offers the best and safest way of commencing what is now called the catheter life under certain circumstances. I will take a case to illustrate what I mean: a short time ago, a gentleman approaching seventy years of age, consulted me in reference to his prostate. Apparently he was in excellent health, but he was going about with a bladder full of urine which reached almost up to his umbilicus. He had never used a catheter, as he could evacuate three or four ounces of urine at a time, which he did at intervals of three or four hours. His prostate was enlarged, the urine was faintly acid, and frequently neutral, and had a specific gravity of 1004-6. Now it appeared to me that

with a helpless bladder of this kind, and with kidneys
working under so great disadvantage it would be extremely
easy to induce that condition of septicæmia which Sir
Andrew Clark has so graphically illustrated. On the
other hand to leave him alone with this distension,
though he was quite unconscious of it, was to him at
that stage of affairs almost equally suicidal. Under
these circumstances I determined to deal with the distended bladder by supra-pubic aspiration, and not to
touch the urethra with the catheter. To commence with,
I drew off four ounces of urine above the pubes with a
fine aspirator needle. I aspirated nine times in a fort-
night, and by gradually increasing the quantity of urine
removed, brought down the bladder to almost normal
dimensions. Elastic pressure over the abdomen was at
the same time exercised by a suitable bandage. The
patient suffered in no respect whatever, and he continued
to void naturally the amount of urine I have already
mentioned. It was interesting to notice that as the
pressure was removed from the urinary apparatus, the
acid re-action became more pronounced, and the specific
gravity rose to 1015 where it remained; under these
altered conditions, I no longer hesitated to use the
catheter daily, which the patient continued to do for himself with advantage and comfort. I thus avoided during
the period of greatest risk in this case, when the urine
was on the verge of alkalinity, and the kidneys hardly
doing anything else than percolate water, any chance of
introducing an important element of decomposition into
the bladder or interfering with a urethra which had
been distorted by prostatic hypertrophy. With acid urine
and a fair specific gravity, with ordinary care there is
comparatively little risk attendant the use of catheters,
where these conditions are absent the danger with

elderly males is by no means an imaginary one; in removing urine either by catheter or the aspirator from the atonic bladder, I feel sure that the employment of abdominal support in the shape of a binder or an elastic bandage is of much advantage. In connection with these remarks on the treatment of some forms of the atonic bladder by aspiration, instead of in the first instance by any attempt to introduce a catheter, I had noted the impunity with which the proceeding had been practised in those instances, where it had followed futile trials to pass a catheter by reason of a large prostate. It seemed rare to find that fever followed this mode of removing urine from the paralysed bladder, though many instances are recorded where it was resorted to for several consecutive days.

Cases of impassable urethral stricture are becoming rarer every day, and this means that a vast improvement has taken place within the last few years in the construction and design of instruments for this purpose. To suppose that one kind of catheter or bougie will answer every purpose, is to suppose that which if not at variance with common sense, is at all events opposed to any practical acquaintance with the pathological anatomy of urethral obstructions.

If after that degree of temporising which supra-pubic aspiration gives in cases of impassable stricture, failure to enter *per viam naturalem* still continues, then I believe Wheelhouse's ingenious proceeding of entering the bladder from the perineum by the urethra is the one that commends itself to our attention. It is safe, and presents no difficulties which a surgeon need fear facing.

I should like to add a word as to the great value of cocaine as a local application in connection with a variety of small operations on the urethra, such as the passing of catheters and bougies, the stretching and dilating of strictures, and the management of various forms of prostatic obstruction where mechanical treatment is a necessity. With a very little trouble and delay, by the injection of a few drops of a 10 per cent. solution of cocaine by means of a suitable instrument, say, for instance, on the face of a contractile stricture, the operation of entering and passing through it with a bougie is in most persons rendered as nearly painless as it is possible. In this way persons will submit to treatment who would otherwise shrink by reason of the horror with which all pain or disagreeable sensation in this part is regarded. I have largely employed it under these circumstances, and am induced to refer to it here because I think its use not sufficiently appreciated in connection with what I would speak of as the minor operations on these parts. To attempt to crush large stones with it, or to use it where the operation occupies some time is merely to bring a useful drug into discredit.